JAVELINA PLACE

The Controversial Face of the Collared Peccary

Edited by

Yvette A. Schnoeker-Shorb

and

Terril L. Shorb

NATIVE WEST PRESS

JAVELINA PLACE:
THE CONTROVERSIAL FACE OF THE COLLARED PECCARY

A Native West Press Book
September 1999

Copyright © 1999
by Native West Press

ISBN 0-9653849-2-6
Library of Congress Catalog
Card Number 99-70525
  1. Collared Peccaries 2. Nature Studies—Peccaries
  3. Animals—Anecdotes 4. Natural History—Southwest

Editors: Yvette A. Schnoeker-Shorb and Terril L. Shorb
Design: Amanda Summers
Cover Photography: Terril L. Shorb
Photography: Terril L. Shorb and Yvette A. Schnoeker-Shorb
Title Page Photograph: Ken Simpson
Digital Imaging: Jerry Chinn
Editorial Support: Tammy Haydon
Production: Triad Associates
Printed by EMI Graphics

For information please contact:
Native West Press
P.O. Box 12227
Prescott, AZ 86304

Manufactured in the United States of America
 Printed on recycled paper

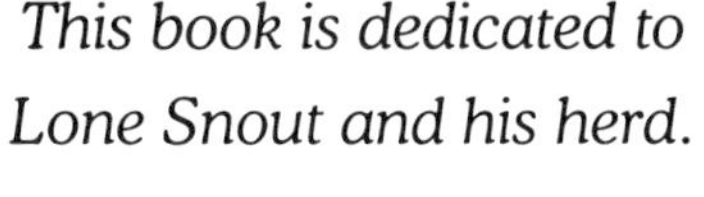

This book is dedicated to
Lone Snout and his herd.

TABLE OF CONTENTS

Introduction

The javelina, also known as the collared peccary, is hooves-down one of the most fascinating wild creatures of the Southwest desert—and one of the most controversial. That the peccary stands out in a land of startling life forms as the giant saguaro cactus, tarantula, and Gila monster is a testament to its status in the wild kingdom. We had barely put down tent stakes in the central Arizona highlands before we began to hear stories about the "outlaw pigs."

Today, many years later, we have had the privilege of spending hundreds of hours in close observation of these wild creatures in urban/forest land interface areas. We have heard many of the tall tales told about them, some of which are reflected within the pages of this book. We have come to know the peccaries as neither pigs nor outlaws.

Straight on—if you are fortunate enough to come upon javelina, they do resemble domestic pigs. There is that pretty, pink snout with its pig-like expressiveness. For a peccary, the snout is the window to the world. Originating in the dense tropics of Central and South America, javelina defensive design favors smell over sight. Their small, rounded ears resemble those of bears more than Old World swine and are effective for alerting the peccaries to danger. When a possible threat is noticed, all the javelina in the herd "freeze" like ballerinas with necks stretched gracefully, hairs bristling to attention, and often a fore or hind leg raised daintily aloft. The only moving part of the peccary may be that athletic pink snout, which, with its dark nostrils, appears as a smiley face reading the world through its "eyes."

Javelina seem to be fashion-conscious creatures, sporting variegated bristles that give them the seasonally-appropriate lighter colors in the summer and more formal darks in the winter, which, of course, are set off by those stunning collars—pale as diamonds and just as flashy—elegantly draped around the shoulders. Javelina are unlike pigs or wild boars in other ways. They have no gall bladder, have stubbed rather than curly tails, possess complex stomachs, and bear only two or occasionally three young. Swine produce litters of 10-20 offspring. The voices of the infant javelina (reds) are distinct and different from a baby pig: a sort of quacking soprano rather than the piglet's wailing squeal.

And what of the stories of the javelina as four-legged marauders who have taken over the territory from the likes of Butch Cassidy and the O.K. Corral desperados? We have heard dramatic tales of javelina menacing Dobermans and bulldozing backyards, of brazened herds entering cozy spaces beneath porches or houses where they behaved like dusty cowboys in town to blow off steam. Could the presence of petite peccaries truly wreak such havoc on the southwestern frontier? Well, like any good tale of the Old West—and the New West—you'll have to saddle up and ride along with us to find out. The book you hold in your hands will take you through some mighty fine territory. Along the trail you're going to meet some humans who have their own stories to tell about their encounters with these critters. We are grateful to all the folks who have contributed both facts and tales, as well as to the javelina who inspired them. We hope by the end of your journey with us, you'll agree that the collared peccary is worth getting to know—and that there is a place for the javelina in our beloved Southwest.

The Editors

Do Javelina Climb Trees?

*An Interview with Eric Gardner of the
Arizona Game and Fish Department*

*Editors' Note: The following was compiled from excerpts of an interview
between Eric Gardner (EG), Region III Field Supervisor for the Arizona Game
& Fish Department, and Native West Press (NWP). Eric facilitated an Urban
Javelina Workshop for the Prescott area public due to increasing incidents of
human/javelina conflicts. In an effort to clear up some general misconceptions
about javelina, we invited him to chat with us.*

NWP: The Southwest is very special because there are species
here that are not found anywhere else in North America or, in
some cases, the world. Unfortunately, as this region becomes
more developed, there is increasing opportunity for conflicts
between humans and wildlife.

EG: There is plenty of opportunity for conflict. Of course, wher-
ever there is an opportunity for conflict, there is opportunity for
resolution.

NWP: When we first moved to Prescott several years ago, before
we had ever seen javelina, we started hearing stories that we
now realize were often misconceptions or tall tales. We had this
vision that these things were the size of hippos with tusks so long
that the beasts would have to turn sideways to get through a
door. When we first actually saw a javelina, we remember
saying, "That little thing is what all the ruckus is about?" Have
you heard javelina described in ways that were amusing or
surprising to you?

EG: Probably the most common tall tale—I don't know if it is a
tall tale—is that they are ferocious animals that intentionally

attack. Anybody familiar with javelina knows that they really don't see very well, and usually what happens is that they are just trying to get out of the area, but they frequently run right towards the thing that they are trying to get away from. So, we get a lot of mistaken reports of actual charges and intentional attacks when, in fact, the animal was more than likely not intending that at all. It was just trying to get away or past or use the only way out. It's like anything. Anybody who has ever, say, been at gunpoint, will describe the gun as being a really, really big black hole right in front of him, and that's all he can tell you about the gun. And, it may have been a very small gun, but the person focuses on that, seeing that really big end of the barrel, such that its characteristics are amplified. I think that sometimes happens when somebody has a close encounter with the javelina. The characteristics are amplified and everything gets bigger—the teeth are bigger, the animal is bigger, and the animal's intent is much nastier than it really is. But, as far as descriptions, people who don't like javelina just think that they're big, ugly, stinky animals that look like bears sometimes. Javelina do kind of, when we see them from the air, look like bears as they lumber down a hillside—just the way they run. Some people really appreciate seeing the javelina and think that they are unique and really interesting looking. Those who don't like them just think that they are ugly, nasty, smelly animals that are not meant to be here—are not natural.

NWP: Do people ever confuse javelina with wild boars?

EG: Sure. We have a lot of that. A lot of people call them wild pigs or wild boars. It is so prevalent that even people who know better, even people in the profession, can be caught saying that. Of course, we all know that they're not. Wild boars, wild pigs, hogs—we hear that quite a bit.

NWP: Another misconception we've heard is that javelina will eat anything including small dogs and children if given the chance. Can you comment about their diet.

EG: Their true diet, if they were left alone and didn't have human influences, is primarily vegetation. They'll eat prickly pear fruits and pads, and they will grub around and eat roots and tubers. They have been known to eat meat, and they will scavenge. Javelina are not known as a predator in the sense that they will not hunt, take down, and kill their food. I think that is one of the main misconceptions. You really don't have javelina out there running around attacking animals to bring them down and eat them. I don't know if people really take it to the second step to think that javelina attack to eat something, but people do believe that javelina will attack to kill as part of their normal behavior. But, you do get the human influence which greatly changes everything. Normal behavior changes quite a bit. And, in the Prescott area, we have javelina that have learned roast beef is a good addition to their diet, so they become a little bit more into the meats and other things.

NWP: So, is it documented that javelina are likely to eat meat if offered that?

EG: Yes. Do you have that report that Cindy Ticer [Arizona Game and Fish Department] wrote? She's got a list in there of plants that javelina will eat and plants that javelina tended not to eat. She also has some stuff in there about what people were feeding the javelina—different casseroles and things. I'm sure that she can provide you a large shopping list of things that javelina will eat if given the opportunity. But I don't really know what they all are because I don't feed the javelina, and most of the people that I talk to deny that they feed the javelina.

NWP: Right— with emphasis on deny. Again, this is coming from what we've heard. Is it true that javelina hate any non-native plant that people grow and will attack it and tear it out by its roots?

EG: I don't know if we could attribute javelina with hate, with that emotion, but I'm not a javelina, so I don't know. What we could say is maybe just the opposite. Javelina love those non-native species because they provide a really unique source of fiber, moisture, nutrients, things that can't easily be found or acquired in a native habitat. What we call those things are "ice cream" plants. There are things that are ice cream plants in the wild when they are in season. The animals will focus specifically on an ice cream plant because it is the cream of the crop, and they've been waiting all year for it to come back. But when people plant tubers and roses and these non-native species, they may offer something that the animals would really enjoy. So, instead of hating these plants, these are something that javelina would love to have in their diet, so they will go out of their way to acquire these plants. Non-native plants offer something that the majority of the native plants don't offer.

NWP: In addition to non-native plants, is moisture an attracting factor for javelina?

EG: I'm sure it is, especially more so in drought years when conditions surrounding the Prescott area are a lot more dry. Food may be more scarce, and water may be less available. Then javelina come into the periphery of the Prescott area and follow the drainages and start to find water from people watering lawns and running water. And, then, lo and behold, right next to the water, there's a deli!

NWP: We were talking to a wildlife rehabilitator who works with javelina down in Green Valley. One of the big problems she sees down there is the development of golf courses, which use a lot of water and grow a lot of green edibles. There are now generations of javelina learning that these golf courses are where the "salad bars" are.

EG: Right.

NWP: We have heard that javelina will chew through a wire fence to get into a garden. Is there evidence of that?

EG: I haven't seen any evidence of them chewing through. Javelina don't really have the type of teeth that are designed to rip or bite through or crush wire. Their teeth are designed more to strip vegetation. But they are very strong animals. I don't think that they actually do so much biting through wire—maybe

breaking it, pushing through it, moving it aside, or digging under the wire. If people build a fence that just isn't strong enough or has gaps that are just a little bit too enticing, javelina will find a way through. Where there's a will, there's a way.

NWP: All right. Are you ready for this one? One rumor has it that javelina have specially adapted dewclaws to help them climb trees to break into hanging bird feeders and to pursue humans or whatever is up there.

EG: Not true. There are lots of rumors about javelina based on their anatomy—like the scent gland on the back was originally described as the navel. They do have a single dewclaw on their rear feet, but javelina are not known to climb trees.

NWP: So they can't tree people? We've heard numerous stories about people being chased up trees by javelina and having to stay up there for hours—even days.

EG: If that were to happen, it would be because the person climbed the tree, and the javelina were foraging below and chose to stay in the area, but they certainly cannot chase up the tree after the person. Javelina will sometimes concentrate on tree areas because there may be food sources at the base of the tree, or it may provide shade or a bedding area. So perhaps somebody may have climbed a tree and didn't feel comfortable leaving because the javelina were still in the area.

NWP: Do you think the javelina knew the person was still there?

EG: Probably not.

NWP: Okay. We've also heard people say that javelina breed like fruit flies. They have litters of 20 young or maybe even more.

EG: That is also not true. Javelina actually have a pretty small area for nurturing young inside of them. Their capabilities are fairly limited as far as how big their litter size can be. I'd say two to three is typical. What happens is that, even though there is a peak season for breeding of javelina, they can breed other times during the year. Herd sizes range on the average around nine or

so animals. And any one of those javelina could have had two to three "reds," so you might have multiple young with the herd. And people may mistake all of those young as having come from a dominant female or one they believe might be the matriarch of the family. In fact, there might be several breeding females in that herd, and they all could have had young. You usually see varying stages of development. You see little reds running with some immature young and some sub-adults, and then adults.

NWP: We have heard javelina referred to as pack animals, much like wolves. Is it true that they travel in big packs so they can ambush prey like dogs or even small humans?

EG: They do travel in herds, but the reasoning would be different than the one that you just described. That would be more of a predatory animal, like wolves, coyotes—that travel in packs in order to hunt and take down prey. Javelina stay in herds for the exact opposite reason—more like why fish would school—to protect themselves from being taken by a predator. They have security in numbers, and they can protect their young that way. Now, because of that herd instinct, they will occasionally attack to protect their young. And where we do see that more frequently is in nature; javelina will defend their herd against coyotes. So, in an urban interface or with a human influence, you may see that same herd instinct go into effect when there is a dog involved. Although a lot of people take their dogs with them for a sense of security from wild animals, in the case of javelina, it may be the one thing that could perhaps instigate something like an attack if the dog decides to get involved with the javelina. I should make this clear. It is typically the dog that instigates that. If the dog gets into the herd of javelina or is trying to attack the javelina, or kill a red, then that herd instinct may take place. But the herd is not running around looking for a dog to take down.

NWP: We want to clear something up. We recently heard a story in which a loose dog chased a herd of javelina until the owner finally took the dog in. A few days later, when the dog again was running loose on the same property, the javelina, who supposedly remembered the dog that upset them, stalked the dog in, well, revenge. Now, is that possible?

EG: No. I would say not. I would say that the javelina were there the first time for a reason, either food, water, or shelter—and they remembered that. Even though the dog was there, it was brought inside, and the javelina then were undisturbed long enough to know (either before the dog showed up or after) that they had located something they liked in the area. Or it could just be that the area is the point between two other locations that the javelina do go to for food, water, or shelter. But a javelina would not purposely go back to seek a potentially harmful encounter unless there is some benefit or reward to it.

NWP: Usually when we hear people describing some sort of negative encounter with the javelina, there was a dog involved. Now, typically, on most county and public lands, people are required to have their dogs on leashes, but what most people confess to us is that Rover was off the leash.

EG: Yes. There would be much less likelihood of an encounter when the dog is not roaming. The owner could then control the dog and make sure that it didn't go in and attempt to attack the herd. Whereas, a roaming dog is going to go out and do what dogs do—sniff around and try to discover everything it possibly can. And if the dog happens to get into a herd of javelina, it is involved in its own little skirmish and probably is not going to hear its owner. The owner's control of the animal is gone. And, then, it is too late at that point. Also, not to mention, the dog might actually run right back to the owner—the herd following the owner home . . .

NWP: Would your recommendation be that, if people are worried about negative encounters because they have a dog, keeping the dog leashed would be best?

EG: Certainly. That is also, in most cases, the law. The average person who goes out and lets the dog off the leash for a run has to realize that, during that time, the dog is under less control. If, in fact, there is a herd of javelina, the ability to control the dog and remove it from that situation is severely reduced.

NWP: Is it true that javelina travel mostly at night because they are naturally sneaky, sly creatures?

EG: Javelina will travel at night, but this occurs primarily in the hot summer months to avoid the heat of the day. Otherwise, javelina usually travel in the early morning and late afternoon. The rumor probably stems from urban areas where the hustle and bustle of daily activity may impair the animals' normal routine and keep them a bit more suppressed and sheltered up and protected. Then, at night, when things quiet down, the vehicles are not on the road, the dogs are back inside, and people aren't working in their driveways and their garages, the urban area becomes the animals' range. The javelina can just go around and seek out things: the dog food and cat food, the casseroles that have been put on the porch, and the tulips that were planted during the day.

NWP: Okay, speaking of planted things, is it true that Arizona Game and Fish Department, sometime in the 1960s, ordered up a bunch of javelina and, during the dead of one night, planted them in the Prescott National Forest?

EG: No. It is not true. Actually, I've heard the rumor. The Arizona Game and Fish Department has always professionally managed this species, and particularly the game species in Arizona. Javelina has been a species of interest for a lot of years. Early on, they were killed in great numbers for commercial purposes. And there were attempts to preserve javelina herds and to encourage their growth. The Game and Fish Department was active in research and ways to ensure that our hunting and harvest promoted healthy javelina populations and did not decimate the herds. We have management plans, and it is all public information as to what we do in transplants and how we are trying to improve any particular species. The closest transplant, that I'm aware of, occurred over near Kaiser Springs along the Santa Maria River and Big Sandy River. Javelina, long before Arizona Game and Fish Department existed, started to show up in Arizona in the 1700s. And, then, over time, you can see how their population has grown and expanded, primarily following the river corridors and drainages. It may be that some of the javelina planted in some of the other areas have affected the Prescott herd, but I would say, actually, the Prescott herd has probably expanded from the south, and the Arizona Game and

Fish transplants have really not had anything to do with the javelina that are currently in the Prescott area.

NWP: Do you think it is possible for humans and javelina to find ways to coexist?

EG: It has to be possible because there really is no other solution. It is possible because it is going on right now, you know. The majority of the people in these urban areas coexist with javelina.

Although it is common that there are home-owner/javelina conflicts, they are really not in great numbers. When a conflict does occur, it is very upsetting to the people involved. It becomes a very important issue to them because there are financial things at stake, damage to property and vegetation. For the most part, the majority of people coexist with javelina. Many people never see them or even know they're here. Others do see them or know they're here and enjoy that. Typically, it takes a specific event when people have incurred damage from javelina and suddenly no longer wish to coexist with them. We try to resolve those conflicts as best as we can. Usually what we have found is that the people who have these encounters are the ones who didn't take any proactive measure. They were unaware of the problem or the potential problem, or they ignored the potential problem and just did things in such a way that the encounter was allowed to occur. They ended up reacting to the problem.

NWP: Could you share some examples of how the Arizona Game and Fish Department is helping people learn to coexist with javelina? For instance, do you have fact sheets?

EG: We sure do. We have got several brochures, such as "Javelina and the Homeowner," "Javelina and Electric Fences," and "Living with Urban Javelina." As a result of homeowner conflicts with javelina, we started the Urban Habitat Committee. We can talk to Planning and Zoning about allowing fencing. We can't make Planning and Zoning not approve a housing development next to a drainage or a good wildlife corridor. What we can do is provide them with information, so that when the homeowners come in, they choose better skirting, or their homeowner association allows fencing to a certain extent. What we have done is written up some general information and provided it to all of the homeowner associations in the area, and we are asking that it be printed up in their newsletters. The agency also has decided to look into using things such as Welcome Wagons and Chambers of Commerce to provide information to newcomers to a community. When people come to the community, they at least are told, "Hey, there are these creatures called javelina, and you may want to plant these types of species and not these others, or you may want to consider fencing, or you may want to consider moving into an area that allows this type of fencing." So, we're trying to find ways to better proactively inform, instead of reacting. We can do things along those lines, but what it all boils down to is that what we really can do is to try to inform and educate.

NWP: Of the calls you get, do you have any sense of the ratio of comments that are about problems as opposed to those from people who are interested in the creatures and don't have problems with javelina?

EG: I don't have a good sense of that. The reason why is because the only people who comment are people who have been impacted, and those are generally then negative comments. It would be unusual for somebody in the community to call the Arizona Game and Fish Department to say, "I just wanted to let you know that I saw a javelina walk through my yard today, and I thought it was really great. And it ate some of my tulips, and I was happy that I was able to feed it."

NWP: You haven't been getting those calls, huh?

EG: No, we don't get those calls. The people who are not impacted don't call us. The people who call us are the ones who are looking for help or assistance, and so the majority of the comments that we receive, I would have to characterize as being negative. But, I don't think it is a true example or ratio of how the community feels.

NWP: Just to conclude, we have had a lot of people ask this question. What good are javelina?

EG: I don't know if I can answer that, not because they are not good, but just because everything has intrinsic value. Javelina are animals that live and thrive in Arizona. They have been here since the 1700s, and they appear to be doing well here and to belong here. They certainly have a place in the ecosystem. So, just like any other animal found in Arizona—what is the value? Javelina are just part of Arizona's heritage.

JAVELINA TALES

Wildlife Expectations
Waiting for Javelina

by Diane Payne

Entertaining overnight guests in Tumacacori is different from other places I've lived. Even though I don't live on a ranch like many of my neighbors, and am wedged between the frontage road and freeway, my guests expect to see wildlife in the yard.

This is a demanding expectation. Most nights Yak, our larger dog, runs to his food dish on the porch and does his menacing deep growl that alerts me to the javelina in the back yard. Barto, the smaller dog, runs inside the house, hoping to avoid any trouble. In the four years we have lived here, the javelina have never bothered the dogs, nor the dogs them. They seem to have worked out their differences peacefully.

The guests are elated when they hear Yak's javelina bark, and they either race to the window or grab a flashlight and sneak outside, thrilled to see the pack of javelina. They'll talk about this event for days. If a night passes without a javelina sighting, the guests are disappointed and unable to sleep, afraid they'll miss Yak's growl and the entire javelina opportunity.

It is unfortunate that most of my guests have good hearing and are light sleepers. Throughout the night I'll see a light go on, and hear them shout from their room, "What was that? What is that noise?"

I tell them not to worry, that it's probably something on the roof. "A javelina?" they'll ask hopefully. It's probably one of those lizards that they get so giddy about when sitting in the yard; yet, that same lizard can create a terrible chaos when found inside the bathtub.

"It sounds like something running across the floor," they yell. I say nothing, hoping they'll go back to sleep. "It's mice," they decide. "I think they're on the bed!"

"You're hearing them eating the adobe walls," I explain, "but they probably haven't made it inside the house, yet," I lie.

"How can you sleep with all that noise?" they ask.

"I could be sleeping now if you'd quit waking me."

My guests are rather selective about which rodents ruin the desert charm and which ones enhance its essence.

It's funny how most mornings the javelina are in the yard when I get up for work, milling around the car, waiting for me to approach, before they lumber off into the field. But, when I have guests, the yard is devoid of javelina. I wonder if it's the strange cars and guests that make them leery. They're cautious creatures for good reasons. Not everyone takes kindly to them being on their property.

Unless there are guests, most nights, the javelina are in the yard, setting off the sensory light, rubbing their backs against the fence, and drinking out of the dogs' water dish. The first time I discovered the javelina was on a Halloween night, and I heard noise in the yard, certain it was my high-school students. I could hear my young daughter's push toys being shoved around, her little cars scraping against the side of the house.

Finally, I toughened up and looked out the window to see which students were in my yard, and I noticed the javelina trampling over my daughter's toys, looking as if they were playing with them. Maybe they were. They returned every night for the rest of the winter.

One night I had a successful javelina visit. Friends came for a cookout, and the food was actually edible. They kept asking where the javelina were. Randomly, I tossed an ear of corn into the compost pile and actually hit a javelina while it was feeding. Both my guests and the javelina enjoyed this. We tossed our ears of corn and watched our party grow in number. That was a party none of us will forget.

But now, both the human and the javelina guests have these wild expectations, which I find hard to fulfill. There are nights when I have both sweet corn and humans, but the javelina don't feel like making an appearance. There are other nights when I'm glad it's just the javelina, dogs, my daughter and I, and the human guests don't make an appearance. And there are other nights when I simply sit outside alone, quietly falling asleep without anyone appearing.

Gangland

by Ann McCarty-Moskus

We put in our forty years. Paid our taxes and our dues. We climbed the corporate ladders and played the games of fools. We breathed in the smog and fought freeway traffic. Raised the kids and paid their tuition. Saw them settled in condos in Malibu. Now it was time to live out our dreams in peace and serenity.

"Let's move to the country," said he. "Away from the crowds and gangs. Someplace where you won't see graffiti covering doors and walls."

"Let's settle in Prescott. Surrounded by trees. Blue skies and clean air," said I.

Yes, this is the life. Instead of the noise of traffic, we wake up to birds in full song. Weather is perfect. Just enough change of season to appreciate the colors and temperatures.

No sooner had the carpet been laid, window coverings hung, and all the boxes unpacked than *they* appeared early one morning. Out of nowhere. Coming from every direction. Quiet. Skulking. Walking softly. Surrounding our house.

A gang. Such a reminder of the gangs we had come to fear in the big cities. All dressed alike to identify their individual gang. Dark gray coats flecked with white. White collars around their shoulders. There were six of them. No wait . .. there's three more. My God, there must be ten of them. Mulling around in our yard. They didn't make a sound. Their very presence was intimidation enough, and they knew it. I peeked through closed blinds, hoping they didn't know I was there. But they knew. They turned toward the house and nodded. Their noses up in the air. They hung around long enough to let me know this was their territory, and I was the intruder. Then they disappeared as quietly and quickly as they had arrived.

John woke up and found me still peeking through the blinds at nothing but forest. I sank into a chair and described the gang I had just observed.

"Ummm," said he. I know the gang you mean. They're tough and they can be mean. Stay away from them. Don't try to make friends. They're very proud. Don't try to give them food. They fend for themselves and even train their young to do the same. They'll steal if they have to, but usually just to survive. Forget about planting a garden and don't put out tulip bulbs."

I watched daily and, just when I had decided they had moved on and wouldn't come around anymore, they appeared again. Out of nowhere as they had before. This time there were at least twenty. I woke John so he could see them.

"Yup, that's them all right! The whole gang. Males, females, and young ones. Man, if you thought biker babes looked tough, get a load of these chicks! Hard to even tell the females from the males. Same hair style, coloring, and all dressed alike. If the females weren't nursing the young right there in front of God and the whole gang, you couldn't tell them apart. Maybe that's part of the plan. Can't tell which is the weaker sex. Or maybe in this gang there is no weaker sex."

"John, they have babies with them. Oh, John, they're so tiny and so cute. Their little coats are lighter. They're reddish brown and have a dark stripe on the back. Maybe it's like the belt colors in karate. They have to earn their colors."

"Yeah, well don't even think about approaching those babies. The parents would kill before they'd let you near them."

"Oh, John, I thought we got away from the fears of gangs when we left the big cities."

"There are gangs everywhere, Darlin'. You just have to choose which ones you're willing to coexist with. Personally, I'll choose these javelina, any day."

The World-Record Boar Of Red Tank Draw

by Judith Fishback

The name is suggestive of gun battles and massacre, but Red Tank Draw is only one of many red sandstone canyons delivering runoff into the Verde River from the southern rim of the Colorado Plateau. We had driven over it half a dozen times on our way to a favorite canyon further on, and as the passenger I could crane my neck at mid-bridge and peer downstream between low rock faces to a flash of green cottonwoods and sycamores. Each time we resolved someday to hike the draw.

It was the morning of the winter solstice, 1978, when Karen, my hiking partner, finally nosed her truck into the greasewood on the eastern side of Red Tank Draw. The day promised to be weakly warm, but in the early morning we were still wearing three layers when we shouldered our rucksacks full of canteens, lunch, and a studious mix of sunbathing equipment.

Leafless sycamores, enormous and elegant, expose their twisted roots on the banks of these washes. Their roots grow bark during dry periods, and describe sensuous pale curves and arcs, often clinging to boulders they have extracted from the earth and brought up into their hearts. On the creek floor the huge brown leaves and bark crackle and snap underfoot until washed away by spring runoff or monsoon flooding. The great trees are stunning and unique, white as salt and smooth, mottled by changing shapes of green and brown where dry bark has fallen away in curling dark chips. Their perfume is sweet and delicious.

As we left the bridge behind and entered a low canyon, we noticed under the sycamores that leaves had been disturbed, and stones freshly overturned. We speculated. What could upturn a boulder this size, and for what reason? We looked for human bootprints, but found none. No cigarette filters, no beer cans, no candy wrappers—no humans since the last flood. So why the little disturbances in the dirt, why the evidence of shallow digging? A bear? Yes. Probably a little desert black bear, we decided, happy to be so near it.

We hiked for twenty or thirty minutes, loving the low sun on our faces, hopping from rock to rock on the floor of the wash and not saying much. All along we noticed the disturbances in the creekbed, and little scrapings under the mesquite and the greasewood along the banks. All signs indicated that whatever had done this clearly was not far ahead, but we were not really expecting to see anything; wildlife in this desert is elusive and wary from decades of being hunted, and it's hard to sneak up on anything when the leaf-litter is so dry.

The wash made a sharp turn, diverted by a vaulting sandstone wall overgrown with chartreuse and orange lichens. Broken shelves of the sheer face were dotted by bonsai vegetation with the verve to send roots into solid rock in the hunt for soil.

Still rockhopping, silently we rounded the curve, and all at once, there before us was a large band of collared peccaries, feeding widely across the wash and up into the vegetation. We froze abruptly as if yoked. "Ohmigod!" I said, under my breath. A native of the Northwest, I had no idea what they were." "Javelina!" Karen whispered, her face glowing. There was the barest scent of musk in the air, as if a skunk had passed by.

Their ballet could not have lasted as long as a minute. It was so well-oiled, so rehearsed, that it was over before we could conceive of what was happening. A boar leapt to an outcrop near the foot of a sycamore. Huge, fierce, he broadsided himself to us and whuffed a tidy mouthful of orders. From his rear the winter sun gave him a long-legged shadow and illumined his bristly hair. He stamped his hoof twice, raising a dusty, bright fog.

The females, without hurrying, seemingly without intent, fell into a half-moon formation, in which the several babies seemed already collected. Their dark little disc-shapes—on tiny whirring legs, trotting single-file in a string behind the boar, each haloed briefly by sunlight, identical as pennies—vanished into the scrub.

The sows followed their young, around the boar on his hillock, into the rising sun. One-by-one they were swallowed up by cholla and catsclaw. The boar did not move from his place until the rest had gone. Then, with a threatening snort, he wheeled and disappeared into his own dust.

Paralyzed with joyful terror, Karen and I both started talking at once. We reran the whole thing, pointing and gasping in wonder and disbelief. "He was enormous!" I said. "How much do you reckon he weighed?" Karen, always conservative in such matters, suggested 150 pounds. I argued for 300. Neither of us could accurately estimate his size, but we agreed he was a huge beast. We relived again and again the awe we had felt looking up at that gigantic, glowing pig perched splendidly on his promontory, and the terrible glare he had beamed at us out of those cold, beady eyes. We praised his glory, and his terrifying ferocity. We were in love with them all for the rest of the day. The babies, we agreed, were completely adorable.

We followed their trail at a distance for awhile. What struck me was how little damage they did. There must have been ten or a dozen, yet the sign would be gone in a day or two. Clearly they had rooted all over the place, but there was no breakage, no real impact. They were like no pigs I knew anything about. I couldn't wait to get my hands on a field guide.

Of course my research Monday at the library tried to tell me my 300-pound boar probably weighed 50. I was sure there was some mistake. In the science department at the university, I flagged down a zoologist, trusting him to support my estimate. "That's the joke on you," he laughed. "These little pigs know how to look big and bad and scary. Everybody's got some 'dangerous javelina' story, especially hunters. They're tough and wily, and you're right, sometimes they're invisible, but they're not very big."

In the twenty-plus years since that solstice morning in Red Tank Draw I have had dozens of javelina encounters. Most recently, I watched a band of them shoot out one at a time from one end of a small metal culvert, like bubbles out of a pipe, and dematerialize on the steep slope of a road cut. Of course, whenever I see them I vividly envision my world-record boar threatening mayhem from his promontory. I've stopped trying to convince other people of the size of him, but I'm still sure he could not have weighed an ounce under 225 pounds.

NURSERY TAILS

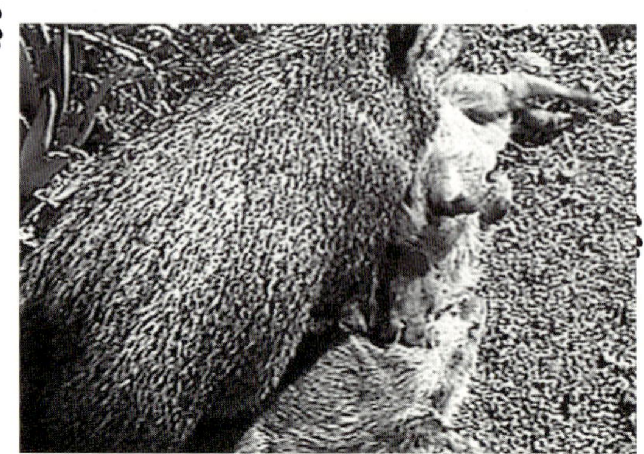

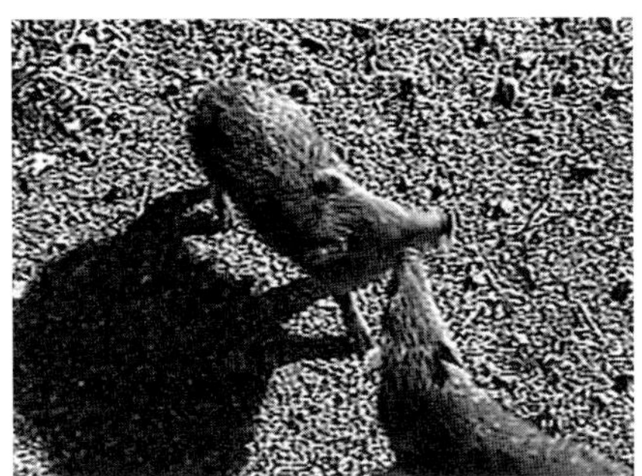

CLOSE ENCOUNTERS
Tracking Javelina

by Graham Wilson

Once, while photographing javelina near Tonto Creek, I spotted a herd of 18 containing three very small ones. I moved downwind 200 yards and blew a varmint call. The herd bunched up, surrounding the three babies, while two larger ones came in my direction. The two separated, circling about 20 yards downwind of me. When they crossed my scent trail, they dashed back to the rest of the herd, and they all took off at a good pace.

Another time, I came upon one javelina chasing another at high speed. They were running in circles, working ever closer to me. Presently, the first one emerged from between two bushes a few feet in front of me. I raised my right foot, preparing to kick it in the head. The javelina slid to a stop right under my foot. I was about to tap it on the ear, but then thought that might not be a good idea. At that instant, it bolted off to my left.

Several years ago, near San Carlos Lake, I was following fresh javelina tracks up a brushy hillside. I knew I was getting very close, as I could smell them up ahead. I crawled forward under some branches as the herd scurried out ahead of me—all of them save one; it was in the process of giving birth. I made a quick 180-degree turn and left her alone.

Another time, in that area, I trailed a herd of 12 to the mouth of a wide shallow cave. Apparently, they were on to me, as they were lined up literally shoulder to shoulder, facing out, alert, with noses twitching. All of a sudden, they caught a good whiff of me and broke into a run down the slope like 12 horses from the starting gate, scattering like a covey of quail.

I live in Wickenburg on the edge of the desert. About once a week a herd of five javelina stop by to nibble on anything I choose to plant and to root over the watering trough.

FISHING FOR JAVELINA
A True Encounter

by Jay Kemp

Eleven o'clock. Time for bed. After opening a window for cross ventilation in the bedroom, I heard the frantic cries from an animal in distress. Recent reports of black bear coming into town prevented me from going out into the dark of night to investigate this awful sound.

The next morning my wife and I returned from an errand in town in time to see two adult javelina milling around on the grass in the front yard across the street. We drove around the corner for a better look. Keeping a vigilant stance, the two critters watched us drive by. Up the street, one of my neighbors was talking to Jack from the TV Cable Co.

After depositing our packages in the house, I returned to the street and tried to determine what was keeping the animals from running off. I heard the same distress call I had heard the night before. My guess was that another javelina was trapped some-where nearby.

Jack had parked across the street. I walked away to the side of his truck away from the pigs, where I could talk to him and, at the same time, keep an eye on the wild critters. Jack told me he had called his dispatcher, Maryann, and she told Cris, who called the police, who in turn were to call Payson Animal Control. Jack thought they should be here any moment.

While we waited, the distressful squeals continued, this time sounding as if they came from the dense foliage beside the house, but the pigs stayed near the fire hydrant on the corner. I noticed an open pipe sticking up about two inches above the ground. It looked to be about eight inches across. I said to Jack, "I'll bet there's a pig in that pipe." He said, "You may be right. Let's find out."

We walked toward the pipe; so did the largest of the two animals, its neck hair standing on end. It grunted a warning. We clapped our hands and both pigs ran a few feet away, then stopped. While the cable man continued to distract them, I took a quick look into the pipe. About two feet down, standing on its hind legs, screaming its little lungs out, was a very upset baby javelina. We backed off quickly. The largest of the two pigs ran to the pipe and stuck its snout down into the opening, as if to console the youngster. Still watching us, it then walked back and joined its mate. They grunted to each other.

Obviously, these animals care about their young. They must have been standing guard all night. Their grunts sounded as if they were discussing something. Maybe it went like this.

"Listen, Myrtal, I don't think these humans are going to hurt Jr. If they do, I'll tear their legs off."

"But, John, I heard the one in the truck call the cops, and you know what that means; those animal control people will come after us again."

"We'll have to take a chance, Myrt. I think if we leave these humans to their own devices, they'll come up with a plan to rescue Jr. I'd do it myself, but Jr. is just out of reach."

About then, Bill, another cable man, drove up, having heard the story from dispatcher Maryann. We explained the situation to him. Still no police came, so we decided to mount a rescue on our own. From the beds of their trucks, the two cable men retrieved a ten-foot extension pole with a hook on the end and a length of rope. One end of the rope was tied to the hook, and the other end was formed into a loose loop.

Bill and I distracted the adult javelina, while Jack dangled the loop down into the pipe.

"See, Myrt, I told you they would come up with something. Yell at Jr. and tell him to grab the rope, then we'll act like were upset and charge them, but don't overdo it."

Oh, for a camera! The scene reminded me of an old cartoon showing a child fishing in a bucket of water. The javelina fisherman bounced the loop up and down a few times into the eight-inch pipe. After several tries—success! Somehow the youngster entangled its front legs with the rope and was lifted out. Jack lowered the crying, twisting baby to the ground a safe distance from the pipe. It wiggled itself free, and the first thing the foot-long javelina did was run back to the open pipe and look into it. Grunting their disapproval, Mom and Dad ran to their wayward child.

I could imagine Papa John shouting, "Jr.! Get away from there! Come on, Mother, let's go home. I'm tired."

They trotted away across the yard and disappeared into the camouflage of the scrub oak at the edge of the forest.

JABALI

by Norah Booth

I am going too fast for the road, enjoying the power of this truck that can easily grab the dirt under its tires, when four javelina, graceful as fish with legs, stream before my eyes. Calculating speed and distance, I know I am going to hit a javelina. Javelina travel in herds from two to 37. Hitting one will not mean anything to the truck. The adult javelina weighs at most 65 pounds, about the average weight of a child. Arizona Game and Fish claims no knowledge of motorists killed by javelina, but 20 road-killed javelina carcasses turn up around Tucson alone each year. Slamming the brakes will not guarantee me anything, except possibly fishtailing on this desert road, providing me less control when I hit. I remove my foot from the accelerator and start tapping the brakes.

The javelina are following a dry arroyo downstream. They live in areas of about a square mile. Although they tolerate the desert with no available water, except by eating cacti, they prefer waterways for the shade and vegetation these support. Disturbed by my engine's noise, the javelina make running leaps almost ten feet off the ground to get away. Five more, including two piglings, enter my path. They all manage to get airborne at once.

I have always considered wild animal sightings good luck. Javelina, however, unsettle me, even though they are "harmless to range, livestock, and people" as one old tract on them states. My ex-mother-in-law once had her fingers severely bitten while hand feeding baby javelina in her yard. She did not know that with each snap of its jaws, a javelina sharpens its interlocking canines.

On one hike I was almost on top of a javelina before either of us knew it. They have poor eyesight and good camouflage. The wind must have been going away from both of us because their musky smell—they have been called "Mexican Musk Hogs"—is unmistakable and offends most humans. Their own sense of smell is keen, and they instinctively run from people. We startled each other that day, and both took off in opposite directions.

Javelina once invaded a porch at a desert house I was caretaking. First one, then two, eventually seven. I stood at the window transfixed by their presence and the rhythmic snuffling noises they made as they uprooted the begonias, tipped over the mums, and stripped the leaves from a tree awaiting planting. I began clapping my hands and scolding like a schoolmarm. "All right, you Pigs, that's enough now. Move along, move along." Like properly reprimanded delinquents, they flowed off the porch, their one-inch brushes of tails sashaying like windshield wipers. The porch garden was ruined.

Call a javelina a pig in front of a hunter in Arizona, and you could quickly get told that javelina are not pigs. "They are *rodents.*"—which, of course, they are not. The species originated in South American and successfully migrated north fairly recently. Missionaries in 1756 made the earliest documentation of Indians eating "jabali." No evidence of javelina bones exists in Arizona before this date. Those that migrated to the Sonoran Desert region are called Collared Peccaries because of the white bristle band around their necks.

The hunter might also explain that javelina (the name comes from the Spanish word *javelin* and refers to their down-turned front teeth) have only two toes and two dewclaws—the little appendage toes like those on a dog—on their front feet and two toes and one dewclaw on their back feet. They leave little heart-shaped tracks. Around here, you get to call javelina "pigs" after you have been informed that they are not pigs. *Tayassu tajacu* is the scientific name, and, truly, they are distant cousins to the domestic pig.

I sit behind the wheel concerned about killing one pig unintentionally. One hunter claims he put a bullet through a javelina's heart with a .357 magnum, and the javelina ran thirty yards before dropping. I know javelina can tolerate a rattlesnake's bite like a bee sting. I am counting on this toughness, those incredible genes, as the truck enters where javelina are crossing. In that moment, my peripheral vision catches javelina on all sides. I am almost through, almost making it, when I hear, *thunk.*

The truck gives a slight shudder. Perhaps it hit the tire, perhaps the side panel. I slam the brakes and fishtail to a stop. I am reluctant to look back. I am no hunter. I have, in fact, what I consider a natural aversion to death, especially when I am somehow involved.

In the road, a pigling sits on its haunches. It just sits there. Finally, a sow comes to the roadway. This is not an adult that has turned back; it is yet another from the side of the arroyo where the javelina have been originating. She whoofs to the baby. The pigling rises unsteadily on all fours, shakes its head, and squealing, runs back to her. I resume my travels, relieved, sloweddown, and oddly shaken by this encounter.

LINDA, I WILL NEVER FORGET YOU!
My Javelina Story

by Nellie R. Montoya

My story begins back in the spring of 1962, on Highway 82 in Santa Cruz County. One morning my father was driving on his seven-hour passenger route, and at a distance he saw something was in the middle of the highway. As the bus approached, he realized it was a little javelina. He stopped the bus and waited for the javelina to move across the highway, but the javelina would not move. One of his passengers stepped out of the bus and moved the little javelina away from the road.

My father slowly drove away, and as he looked back through the passenger mirror, he saw the little javelina moving toward the road, then sitting in the middle of the road again. He stopped once more, and one of his passengers stepped out and brought the javelina inside the bus. My father made sure the little javelina was all right. The bus had a large compartment in the back, so he arranged a safe area for the javelina to rest until he could figure out a solution. When my father arrived at his destination, he gave the javelina water and food. When my father completed his route, it was sunset. That was the night I met the little javelina.

I was six years old and had never seen a javelina before. It was a female, and she wanted us to lift her in our arms. She had corn flakes for dinner and was happy. The next morning my father and I took the javelina to the veterinarian. He examined her and told us she looked very healthy. The veterinarian also said that this young javelina was about four months old, weighed 18 pounds, and was *domesticated*. It was obvious that the little shoat had been bathed because she was very clean. She was very reliant on humans. The veterinarian told us that she probably would not be able to survive on her own. We left the clinic with her. For days and days my parents waited for a notice in the newspaper for a "Lost Baby Javelina," but we never heard from her previous owners.

When we arrived home, my father and my mother decided to adopt the young javelina. Now, we had to find a name to call her. We sat down and watched her play, and we decided to name her "Linda" (*Pretty* in English). Linda was a lot of fun. She followed me everywhere. I had a pet Chihuahua whose name was Pichiquin. Linda became friends with him. We also learned that Linda was very smart and understood us very well when we spoke to her.

Linda practically lived inside the house. When my mother washed the laundry and went outside to hang clothes, Linda followed her and played on the patio. She would always stay very close to my mother and rubbed my mom's ankles with her nose. Sometimes when my mother went outside, she would leave Linda inside. Linda would squeal and squeal because she did not want to be alone.

Linda would go with us almost everywhere. She liked to ride in the car and enjoyed sitting on my lap. When we watched television, she would start running from a distance and pick up speed, so she would be able to jump up onto the sofa. She would rest her head in our laps and rub her nose on our laps, legs, ankles, or arms.

Linda had a unique way of eating her meals. Her bowl was in the kitchen, and she would drag her food from the bowl all the way across the kitchen floor. Then Linda would begin her way back to the bowl, eating all the food she had spilled. One of her favorite treats was corn flakes with milk.

Also, Linda enjoyed when my mother gave her a bath; she did not want to come out of the tub.

Everyday when I came home from school, she was waiting for me. Linda, Pichiquin, and I would chase each other around and around. After a while, we were tired and needed a little nap. Linda slept in my room, and I always gave her a hug before

I went to sleep.

As time went by, Linda began to grow and grow. When we were watching television, Linda was no longer able to jump onto the sofa. We had to lift her so that she could sit with us. Her food portions increased; for example, a bowl of feed and corn flakes was not enough anymore. Her bowl became a bucket, and her squeals became louder and louder. By this time, our javelina probably weighed about 35 pounds.

Summer came and she enjoyed bathing outside in the sprinkler. She loved to roll over on the lawn. Linda's growth made her stronger and stronger. She still loved to rub her nose on our laps, legs, ankles, and arms. But her strength was increasing, and now she would give us bruises.

Fall arrived and Linda's weight went from 35 pounds to 50 pounds. I was at school, and my father worked long hours every day, so my mother was the only one that was able to take care of Linda during the day. We lived in a small two-bedroom house with a small patio. My father always said that a pen was not a place for Linda to be. He said that Linda should be able to move around.

A couple of months went by. One night, like every night, I gave Linda a hug and said goodnight to her. When morning arrived and I got up to go to school, I could not find Linda anywhere in the house. I asked my mother about Linda, and she told me that my dad had taken her to the veterinarian for a checkup. I went off to school, wondering if Linda was all right. I could not wait for the school day to end. I arrived home and Linda was not there. Then my mother broke the news to me.

She told me that Linda would not be coming home. My father had found a ranch with other animals and where Linda would be much happier. I remember I broke down in tears because I knew I would no longer have Linda with me. However, as time went on, I began to understand how Linda would be more comfortable at a ranch where she could run and play. I never saw Linda again; her new home was far from the area where we lived. Now, after over forty years, I still remember the fun times Linda and I shared.

WHAT'S CUTE, RED, AND HAS NEEDLE-SHARP TEETH?

An Interview with Sue Simpson of Simpson's Wildlife Sanctuary

Editors' Note: Sue and Ken Simpson are wildlife rehabilitators in Green Valley, Arizona. This interview took place at Simpson's Wildlife Sanctuary in the Sonoran Desert, amid mesquite trees and a compound of enclosures in which various wild creatures live while they heal from mostly human-caused misfortune. The interview was conducted between Sue Simpson (SS) and Native West Press (NWP).

NWP: How did you come to this work of rehabilitating javelina?

SS: In 1979, while I was working for a veterinarian, I had an opportunity to work with an infant badger. We began receiving many different species, including javelina, hawks, and owls. Within a few years we were also doing migration studies and more rehabilitation with raptors. I went to see a raptor specialist, one of these guys who uses an ultra-light to train birds for migration. I mentioned to him that I thought I would specialize in raptors, but I'd like to find a way of making money while doing some of this work. He said, "If you are going to research something and you want to make some money at it, pick something that *stinks*." So, of course, what's the first thing that comes to mind—a skunk, right? I didn't know if I even wanted to rehabilitate skunks. With true research, you are really involved, and they are really in your face. Then I thought, well, maybe vultures would be interesting. But we have rehabilitated vultures in the past, and you know the vulture's defense mechanism is to vomit when threatened. So, no, we're not doing vultures. Then, I decided to look into javelina because, with javelina, there is a tremendous amount of knowledge that we are lacking. This was in 1983. When we started working with javelina, we found them to be truly enjoyable. They are fascinating creatures. They are as bright as can be. They put a puppy to shame. And they are beautiful, especially as infants. Also, working with and studying javelina was a research project which would contribute to our javelina knowledge bank. Dr. Lyle Sowls, at the University of Arizona, had written a proposal for the best way to rehabilitate javelina. He wrote this proposal in conjunction with Gerald Day, who is another well-known javelina specialist, and a team of biologists from the Arizona Game and Fish Department. They came up with a plan. We took that plan, plus $2,000 of our own, and put it into action.

NWP: How did things go in those early years of your javelina rehabilitation project?

SS: Once we started on the project, we found two things happening. One, if we did this project over a number of years and found that this was an unsuccessful way to rehabilitate javelina, we might be closing the door on ourselves and on all wildlife rehabilitation or, at least, javelina rehabilitation. If rehabilitation produces nuisance animals, then the government must look at the program and decide whether it is functional or not. Second, we found that people pick up infant animals. We can't help it. It's human nature. A baby javelina is adorable. You saw that little one out back. Your first inclination is to want to pick it up and hold it and love it. But the first time you go for that baby javelina, it will try to bite you—and it will do a good job of it. They have sharp little teeth, like little needles. However, once baby javelina have decided you are family, they are adopted—they are *yours*. At this point, only *humans* are family. Other javelina are just strange-smelling, spooky creatures. At about four months old, they start becoming very aggressive toward anything that comes between their family and themselves. All we can do is provide a better situation for the young javelina. When we started into this project, we had two goals: to see how rehabilitating javelina can work realistically and to provide a place for people to take javelina where they have a better chance of being prepared for release. We have been quite successful with this endeavor.

NWP: And how about the release of the javelina back into the wild? How is that going?

SS: We are very fortunate to have Scott Richardson, an Arizona Game and Fish biologist, who is studying every release we make. Each animal is monitored by use of a radio collar, and students go out on a regular basis to observe how the javelina are doing, what helps, and what is not helpful. Are the released javelina able to find food? What kind of habitat is best for them after rehabilitation? A wild animal has an instinct to go search for certain foods. Now, I could introduce them to those foods here, but the javelina still have a foundation of domestic pig food because we want to make sure that they are healthy and fat when they leave here. We want them fat and strong enough so that they can get through an initial couple of weeks while they are setting up their territory, looking for food, looking for a water source, and for a place to hide. They are going to lose a lot of weight at this time. This portion of the project is controlled by Scott and Arizona Game and Fish. We feel very confident of Scott's abilities.

NWP: What kind of people bring in baby javelina to be rehabilitated?

SS: We've had hunters come in, all dressed up in their camouflage clothing, with guns dangling everywhere. They have this little, bitty baby javelina in their huge hands. These great, big guys—they're always at least eight feet tall—have this baby javelina tucked up under their chin. They ask what we are going to do with the baby. If I had a picture of that, I'd love to put that in your book. People often assume that the hunter is the bad guy because they'll say, "He shot the mother!" Well, the fact is that adult javelina out in the wild are difficult to identify by gender. The male's testicles are held up much higher on the abdomen than most animals, and, consequently, you do not see them. You cannot go by size because very often the female may be as big or bigger than the male, depending upon how much food is available. An infant doesn't always follow the mother because an infant will follow any adult in that herd. It is a situation where you don't know. You are often not even going to see the infant. I will hear the hunter say, "I shot the animal and then I saw the baby," and the hunter will be in massive distress. I have had hunters tell me they will never hunt javelina again because they found the infant and realized what they had done.

NWP: What is the status of javelina as a viable, wild species?

SS: Javelina are in great abundance right now. There is no worry about the species becoming endangered, unless you have

something that is totally devastating. *That can happen to javelina.* Javelina live in pockets, and there are basically only three genetic lines of them in the U.S. Their range has slowly expanded from Mexico to the north, but they have only been here in southern Arizona since the 1700s. This expansion of their home range occurs when a herd of maybe 20-25 animals lives in an area that may not have enough food, water, and shelter to support them. So a few will break away and move farther north. They become the nucleus of a new family. This nucleus will also grow to a herd of 20-25, then more will break away.

NWP: So with limited genetic lines, does this endanger their long-term viability in the wild?

SS: It could very much so. If we find there is some sort of disease that is genetically connected, it could destroy all new herds. For instance, we had a mutated form of distemper that entered into some southern Arizona javelina populations. It wiped out whole herds, whole pockets of javelina. That happened a few years ago, and it could have been much worse than it was.

NWP: So, even though there are currently lots of javelina, it is not known if this is a healthy, durable population?

SS: This is the kind of thing wildlife rehabilitators really need to be looking at and studying. For example, we have a good population of deer, a solid population. You hear people saying, "Oh, the deer population is greater now than it was in 1492 before Columbus came to America." This is true. Yet it was a different population. Back then there were whole populations of creatures who had undergone genetic straining, and it culled out the weaker lines. Those left to breed were the strongest, and we had this very secure population of deer. The population we see now is more susceptible to diseases than the population in 1492. So even though our numbers are larger, we could see that whole population drop down to nothing very rapidly by having some new disease come in.

NWP: Where do these diseases come from?

SS: The world is changing rapidly. We are seeing new diseases through mutations all the time because of what humans are releasing into the atmosphere and the food chain. We are causing this. If we are going to cause this kind of breakdown in the genetic makeup, in the tolerance of our creatures, then we must look at what we can do to ensure their existence if something devastating happens to a population. If we don't try this javelina rehabilitation project now, and a whole species is wiped out, then we are back where we have been with the Mass Bobwhite quail, the wolves, whales, snails, and fish species that are endangered. We have this small group of animals, very small genetic package, that we are trying to start whole populations of creatures from. I'd rather try it while we still have lots of animals.

NWP: Due to loss of habitat and other factors, javelina continue to push northward. They are in the central Arizona highlands now. But aren't these creatures pushing the limits of their natural tolerances? They are, after all, tropical creatures.

SS: That's true. They are still being pushed northward. But here is the other side of it. We are actually producing a whole population of javelina that are adapting to human presence. The javelina is a prey animal that has moved in around humans. Javelina are docile unless provoked, and they have to be provoked pretty heartily before they'll do anything. Because of their relative blindness, the one way you can trigger problems is by surprising them at close range. But if you have a little common sense, you can handle the situation well. Just skirt around them and they will walk away. They really don't like to be around humans. But if they learn to be comfortable with humans, and are accustomed to human scent, having smelled it over and over, they can reason that the human over there is not a threat— javelina are very intelligent creatures.

NWP: Are you saying that our increasing human presence in the javelina habitat is actually changing javelina behavior and maybe even their natural tolerance for conditions in this extreme climate?

SS: Yes. Food and water in the desert dictate how everything acts. Humans put in golf courses with pretty green grass. We put in ponds and drip systems, and we put in lovely bright red and

yellow flowers. Then we say, "Why are these javelina coming into my yard? Why are we seeing them in the middle of the city?" Because we didn't shut down their roadways. The roadways are the washes. Javelina have been running those roadways for centuries. How can we all of a sudden turn around and say, "Sorry, you can't come here anymore."? We are finding whole populations of javelina and other species that are so adapted to humans that this is going to be a major focus within ten years. We are no longer going to have to worry only about rehabilitating infant javelina that are abandoned by wild ones. We are going to have to learn how to rehabilitate those who have become so imprinted that this is their natural way of focusing on life. I really believe this will have to be the next focus for wildlife rehabilitators.

NWP: Can these animals be rehabilitated to be less adapted?

SS: Well, we humans need to be re-trained. I think we will have some success if we can take the javelina away from the human element and re-release them into an area where there is no human influence. But, once again, we come back to the fact that this is a bright creature. A javelina can learn processes, figure out how to find humans. And the javelina will wander around and *will find humans.* Then there is this other whole side of the issue. The javelina that are habituated to human beings are also habituated to more moisture and more food than wild javelina in the desert. This is a huge factor that not many people are looking at right now.

NWP: We are destroying prime habitat all over the Southwest to make subdivisions with golf courses. How are these, specifically, affecting such wild creatures as javelina?

SS: It would be a wonderful study to look at two different herds of javelina, one which has been totally habituated by humans and has been living on a golf course for a while compared to a captured wild herd—to see what their physical systems are like. I think what we will find is that the habituated animals are going to have a much greater need for water and different types of nutrition than the ones in the wild. We have done this to them. Every new generation is more "successful" [at human-habituated living]. The mother comes directly from the wild, raises young on the golf course, dies off; now, those young grow up and raise their own. If this happens for two or three generations, you cannot take that herd back out in the desert and expect them to survive. They won't be able to survive. They desperately need that [golf course] water, and they desperately need that food. The new generations no longer know how to look for food other than green grass. Where are you going to find green grass in the middle of the desert, except for a couple of months out of the year? Even then, a lot of the desert grasses are not high in nutrition. We have some real problems now. Some studies have been conducted on herds of habituated javelina. The researchers got some good results because they put a lot of time in, going out at night, constantly monitoring the javelina. Personally, I would rather see that much time and effort put into changing and educating people.

NWP: Is there anything people can do to help the javelina?

SS: The best thing you can do is to be a responsible homeowner. We have all kinds of laws that protect us. You have to be responsible if you are going to move into an area where there is a lot of wildlife. Do your research. My biggest thing I tell people when I give programs in Green Valley is to not buy a house there until you go to the Desert Museum and see what those animals are. You see where they live because the Desert Museum imitates their natural habitats. So you can look in your back yard and say to yourself, "I've got mountain lion habitat right here, and that mountain comes right down to my fence, and I've got a pussy cat. Is this going to be a good place to live?" People will have feed out for quail on one end of the yard and little hamburger balls for the roadrunners on the other. Now, the quail start nesting, and the young hatch out of their eggs. All of a sudden the people see the roadrunner going to get the eggs, and they are all mad. They will call us up and say, "Come and get rid of this nasty old roadrunner!" When I ask why, they say, "He doesn't like hamburger balls, anymore." And I say, "Hamburger balls have no legs to run, and it is more fun to chase those baby quail. Remember, roadrunners are marvelous, wonderful creatures, yet they are predators. Yes, he can come to your door and knock on it with his little beak, but he can be very aggressive." So people move in and feed him hamburger balls, and they love him. But the first time they see him eat a baby quail, they hate him with a passion, and then they want him dead. Come on, people, let's step back and think about this first. Let's respect Nature's way.

NWP: Do you have other recommendations for people to help them to be more responsible to the wild creatures whose home range the people have moved into?

SS: There are ways of interacting very naturally. You can attract birds into your yard with a certain type of vegetation. You can attract certain mammals, and you can repel certain types of mammals. Fences—if you want to keep certain animals out, then you must research the method. Game and Fish offices all over the country are getting heavily into education. The education departments from most of these offices are wonderful and are well prepared to tell you how to coexist with wildlife in your particular area. Again, it is important to understand your own area. If you live in an area that has an abundance of food, you don't have to worry as much, but if you live in an area like the desert, there is not an abundance of food of any sort.

NWP: What if people have javelina naturally occurring in their immediate yard area?

SS: You saw all that shade cloth we have? You might consider that around a garden because, with that, as far as the javelina are concerned, you aren't there. They know your scent, and they are comfortable with it, but you have somehow not entered their reality. That shade cloth keeps us out of sight. The other thing is that some wild creatures aren't so upset if they can smell you, or if you have four legs—like when you are on a horse. You get off the horse, and now you are two-legged. That is terrifying to a wild creature. As long as you are up on a horse, the smell is still human, but it is mingled with an animal that is four-legged. So the javelina is a little confused and may respond by just leaving. All enclosures at our rehabilitation center have shade cloth on the lower half. None of the animals see us walking around. They may see us from the waist up, but they don't know whether we are four-legged or not. Our odor is there—we can't do anything about the odor, and there is a certain amount of sound pollution they are getting from human voices, but they are going to have to deal with that just about anywhere they go anymore.

NWP: You work hard at limiting human contact with javelina?

SS: Yes. Some come in as full-grown adults. Normally what they will do is just ram against the side of the cage until they kill themselves. Or sometimes they will cause a new injury, or they will rip apart the fence. Often we get a javelina with a broken leg, and we put a cast on it, and the javelina will go tearing into that cast. They just don't understand; it doesn't make any sense to them. If it is a major injury, my recommendation is euthanasia. Some rehabilitators will tell you differently, but I'll tell you this up front. If a javelina comes in to us, and we don't feel that it can be healed well enough to survive in the wild and reproduce, we will put it down. If we would have left it alone, it would have died and gone into the food chain. When we entered into wildlife rehabilitation, we had to figure out our responsibility. We decided we would not interfere with animals unless we could do a little better than what nature could. If I have to cut off a leg of a javelina that's been injured by a car, how is a three-legged javelina going to survive in the wild? Some little creatures you can do that with. We've released a number of one-legged roadrunners, but only ones that had a good-sized stump left. They kind of look like they are running with a skateboard. We release them right here where they have plenty of opportunity to find food and shelter. They know the area, and they are not too large where they would be a problem. But a three-legged javelina? I don't think so. What right do I have to take that javelina in, cut off the leg, heal it up, and release it so that something else can kill it right away? This is not a game.

NWP: So, you don't have "education" javelina—creatures that cannot be released back into the wild due to the seriousness of their injuries and who are kept for public educational purposes?

SS: Not really. If I have a cute little javelina, it's just too cute and people can't get beyond that cuteness. People ask if it is a pet or if you feed it, and, even though you say no, they smile at you, walk out the door, and think, "I'd a fed that thing in a second."

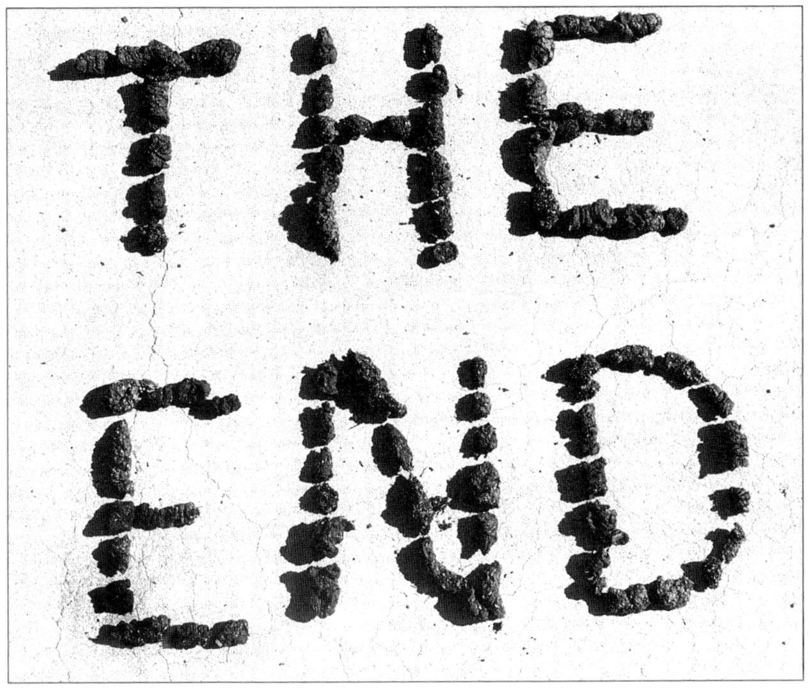

(Produced by the Peccaries of the Prescott National Forest)

Also from NATIVE WEST PRESS ...

Least Loved Beasts of the Really Wild West: A Tribute

Least Loved Beasts of the Really Wild West: A Tribute is a wonderful book. It is filled with insight, understanding, compassion, and an occasional and necessary dash of humor. This book helps us to transcend our often deeply imbedded biases and arrogant and sometimes foolish presumptions about some of our less favored brethren. I highly recommend this book to all those who harbor a deep affection and affinity for the natural world.

> —Stephen R. Kellert, Ph.D.
> Yale University School of Forestry and Environmental Studies
> Author of *Kinship to Mastery: Biophilia in Human Evolution and Development* and *The Value of Life*

Like the creatures it reveres, this little book must grow on you. Stick it in your daypack and read it in the woods after some multi-legged or legless critter has triggered your personal squeamishness. Use it to grow a new set of eyes.

> —Harley G. Shaw, Wildlife Biologist, Author of *Soul Among Lions*

The Spiders and Spirits of Petunia Manor

For the love of spiders! A true story of the magical place between tribulation and transformation. *The Spiders and Spirits of Petunia Manor* is a celebration of the natural world and its healing effects—if humans will but humble themselves to learn from the life that dwells in their own backyards and, if they're lucky, hangs in the overlooked, dark corners within their own homes.

For more information or to order this book, please contact us at:
Native West Press
P.O. Box 12227
Prescott, AZ 86304

To order *Least Loved Beasts of the Really Wild West: A Tribute*, please send $8.95 plus $1.50 for shipping and handling ($8.95 plus $3.00 for Priority Mail) to the address above.

To order *The Spiders and Spirits of Petunia Manor*, please send $6.95 plus $1.50 for shipping and handling ($6.95 plus $3.00 for Priority Mail) to the address above.